Ruffe

T0191095

This book is dedicated to the spiky little freshwater ruffian known better as the ruffe, pope, 'tommy ruffe' and other local names.

A fascinating little fish, the ruffe is long overdue a book all of its own. Much loved by many anglers, ruffe can also be problematic when introduced beyond their native range.

Scientist, author and broadcaster **Professor Mark Everard** details fascinating aspects of the biology, angling and wider contributions to society of the ruffe.

MARK EVERARD

Ruffe
The Spiky Little Freshwater Ruffian

CRC Press
Taylor & Francis Group
Boca Raton London

CRC Press is an imprint of the
Taylor & Francis Group, an **informa** business

First edition published 2023
by CRC Press
6000 Broken Sound Parkway NW, Suite 300, Boca Raton, FL 33487–2742

and by CRC Press
4 Park Square, Milton Park, Abingdon, Oxon, OX14 4RN

CRC Press is an imprint of Taylor & Francis Group, LLC

Library of Congress Cataloging-in-Publication Data
A catalog record for this book has been requested

ISBN: 978-1-032-31732-8 (hbk)
ISBN: 978-1-032-31731-1 (pbk)
ISBN: 978-1-003-31103-4 (ebk)

DOI: 10.1201/9781003311034

Typeset in Joanna MT
by Apex CoVantage, LLC

Contents

One

Does a little freshwater fish like a ruffe deserve a book all of its own? As far as the larger publishers are concerned, that's a definitive commercial NO! Fortunately, as I lack a business brain and consider the merits of this quirky little fish in their own right, my answer is a definite YES!

So, welcome to this book dedicated solely to the commonly overlooked ruffe. This small fish, in common with smaller mahseer, simply does not believe that it is small! The ruffe has something of a localised distribution in Britain. Despite that, it is nonetheless familiar to many anglers and others interested in freshwater wildlife.

Going by the Latin name *Gymnocephalus cernua*, ruffe are also known by various names including 'pope'. They are locally common in rivers and still waters across a broad native range spanning Eurasia (Europe and Asia). Ruffe have also been introduced well beyond this native range, both more widely in Eurasia and into North America. Like many other organisms that become established beyond the localities and ecosystems within which they co-evolved, ruffe can cause major eco-logical disruption in these new regions.

Ruffe are not large fish. Though recorded up to 25 centimetres (almost 10 inches) long and weighing up to 400 grammes (a little over 14 ounces), ruffe generally only reach half that length and up to a third that weight in British waters.

The ruffe's body colour is highly variable, depending on the environment. In highly turbid canals and other murky

DOI: 10.1201/9781003311034-1

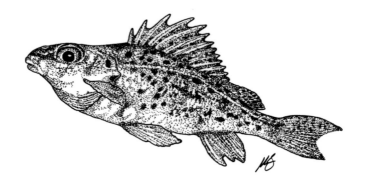

waters, they can be largely silvery or grey with darker blotches. However, ruffe from clearer water take on an olive- to golden-brown or khaki hue on the back and flanks, over- lain with numerous, small and irregular dark blotches, fading to yellowish-white on the underside.

In my book (this one as it happens), ruffe are more than deserving of a book all of their own, regardless of narrow commercial considerations. They are, after all, fascinating little creatures whether loved (by many an angler) or loathed (by conservationists worried about their often-severe ecological impacts as alien invasive species).

Two

The ruffe known to British anglers and naturalists is most properly referred to as the Eurasian ruffe, going by the Latin name *Gymnocephalus cernua* (Linnaeus, 1758). However, across Europe and Asia, there are five species within the genus *Gymnocephalus*, all of them variously known as ruffe.

These various ruffe species are widespread in the fresh waters of Europe and eastwards into Asia. Only one species, the Eurasian ruffe, occurs in British waters and is the species that will preoccupy us in this book. So, where you see the word 'ruffe' without other qualification, this is the familiar species we are talking about.

RUFFE TAXONOMY

Ruffe are part of the class of ray-finned fishes (Actinopterygii), within which they are classified in the order of perch-like fishes (order Perciformes).

Within this broad order of perch-like fishes, ruffe sit within the perch family (Percidae). The Percidae includes 239 species, from 11 genera, found in fresh and brackish waters across the northern hemisphere. The perches have two dorsal fins, which may be separate or else narrowly joined.

DOI: 10.1201/9781003311034-2

Despite the resemblance of ruffe to small perch, the species are quite distinct. However, in his book *British Fresh-Water Fishes*, the Reverend W. Houghton (rightly) records:

The Ruffe, resembling in outward characters and markings both the Perch and the Gudgeon, has sometimes, but erroneously, been considered a hybrid between the two; although there is actual proof that closely allied species do, occasionally and perhaps not infrequently, cross, this is not true of fishes so distantly related as the Perch and the Gudgeon.

As already noted, there are five species of ruffe across their natural Eurasian distribution. These five species are briefly described in the Box overleaf.

The five ruffe species of the genus *Gymnocephalus* are:

- The Eurasian ruffe *Gymnocephalus cernua* (Linnaeus, 1758) is the species that is the subject of this little book. It is the ruffe that is familiar to anglers, naturalists and conservationists in Britain. It is also the one that has been introduced into North America, with associated problems.
- Donets ruffe *Gymnocephalus acerina* (Gmelin, 1789) is a rare fish found in drainage basins of the northern Black Sea and Sea of Azov in Dniester, South Bug, Dniepr (up to Belarus), Don rivers and Kuban drainages.
- *Gymnocephalus ambriaelacus* (Geiger & Schliewen, 2010) is endemic to Lake Ammersee in the upper Danube basin, southern Germany.
- The Danube ruffe *Gymnocephalus baloni* (Holcík & Hensel, 1974) is, as the name suggests, a species found in the Danube river system and the Dniepr catchment (which rises in the Valdai Hills in Russia flowing through Belarus and Ukraine to drain into the Black Sea). It is also believed to be present in the Dniestr catchment (running through Ukraine and Moldova before discharging into the Black Sea).
- The schraetzer *Gymnocephalus schraetser* (Linnaeus, 1758), or 'striped ruffe', is native to the Danube basin. The schraetzer is distinguished from the Eurasian ruffe by its more elongate form, longer snout, and a different number of spines and rays in the two dorsal fins.

Having acknowledged these various ruffe species, let's get back to the species of interest here: the Eurasian ruffe *Gymnocephalus cernua* (Linnaeus, 1758). More commonly, we refer to it simply as the 'ruffe'.

The Eurasian ruffe occurs naturally throughout Europe in the Caspian, Black, Baltic and North Sea basins, including in Great Britain, where it is the only native species, and as far northwards as approximately 69°N in Scandinavia. It is also found in Asia in the Aral Sea basin and in the Arctic Ocean basin eastward to the Kolyma drainage.

The Eurasian ruffe is also now found in a number of other countries as a result of introductions. As we will see, ruffe – as indeed many other plant and animal species – can be problematic when introduced and established beyond their native range into ecosystems within which they have not co-evolved.

KEY FEATURES OF THE RUFFE

The body shape of the ruffe is similar to that of the perch, approximately oval in cross section and elongated, tapering from a front half but with a high back on which the dorsal fins are placed. The flanks are fully covered by small scales that are slightly rough to the touch.

There is a continuous lateral line extending the length of the body, along which are 33–42 lateral line scales.

As Izaak Walton wrote of the ruffe in *The Compleat Angler*:

> There is also another Fish called a Pope, and by some a Ruffe,
> a fish that is not known to be in some Rivers; it is much like the
> Pearch for his shape, and taken to be better than the Pearch, but it
> will not grow to be bigger than a Gudgion; he is an excellent Fish.

They are not large fish. Ruffe have been recorded as growing up to a length of 25 centimetres (almost 10 inches), though mostly the ruffe encountered in British waters reach only half that length.

The maximum recorded weight of a ruffe is 400 grammes (a little over 14 ounces), although, once again, this is not from British waters, where they reach a maximum of only around a third of this weight.

The body colour of the ruffe is highly variable, depending on the environment. In highly turbid canals and other murky waters, low light penetration means that the camouflaging colour is less critical. Consequently, the general colouration of ruffe in these opaque waters tends to fade largely to silvery or grey overlain with darker blotches. By contrast, ruffe from clearer waters tend to be far more brightly and variably coloured. Body colour here can range from olive-brown to golden-brown or khaki on the back and flanks overlain with numerous, small and irregular dark blotches. The underside of the fish fades to yellowish-white.

Ruffe have two dorsal fins. The front dorsal fin is spined, whilst the rear dorsal fin is supported only by soft rays. Unlike the dorsal fins of the perch, in which the two dorsal fins are clearly separated, those of the ruffe are fused together and connected without a notch. The two dorsal fins of the ruffe are cumulatively supported by 11–19 spines and 11–16 soft rays (XI-XIX/11–16 in scientific notation). There are dark spots on the membranes between the spines and soft rays of the dorsal fins.

The conjoined dorsal fins are a key feature distinguishing ruffe from small perch. However, perch are also characterised by strong vertical stripes on their flanks.

Beneath and to the rear of the ruffe's body, the anal fin has two spines and 5–6 soft rays (II/5–6). The caudal (tail) fin has 16–17 soft rays. The tail fin and the front dorsal fin carry the body colouration, including irregular blotches and spots, the front, spined dorsal fin lacking the dark spot at its rear, as commonly observed in perch. The other fins of the ruffe tend to be paler. The pectoral fins are supported by 13–17 rays.

The head of the ruffe is rounded, scaleless and coloured the same as the body. The relatively small mouth is positioned centrally, though slightly down-turned (inferior) at the front of the head. It is armed with an even row of tiny teeth in the jaw. Ruffe also have a small patch of vomerine teeth (also known as palatine teeth) on a plate in the roof of the mouth. In common with other members of the perch family, ruffe have no barbels (sensory whiskers) around the mouth.

The ruffe's glassy eyes are relatively large. Kenneth Mansfield notes in his 1958 book *Small Fry and Bait Fish: How to Catch Them*:

> *An unusual feature of the colour scheme of this fish is the purple eye.*

The head is also interspersed with conspicuous slime-filled sensory canals just beneath the skin. These are particularly noticeable below the eye and on the preoperculum (a plate overlying the operculum, or gill cover). As described by Alwyne Wheeler in his 1969 book *The Fishes of the British Isles and North-West Europe*:

> *A series of large sensory canals just beneath the skin in the head are particularly noticeable below the eye and on the pre-operculum.*

Within these sensory canals as well as those of the lateral line, neuromast cells enable ruffe to detect fine vibrations in the water, aiding them to find food and avoid predators.

The operculum, comprising a series of bones supporting the facial structure and extending to cover and protect the gills, is spiny with a sharp spine at the rear end. This feature is also shared with the perch.

However, unlike the perch, the rear of the ruffe's preoperculum is also scalloped and spinous. As described by Kenneth Mansfield in his 1958 book *Small Fry and Bait Fish: How to Catch Them*:

> The plates of the forward gill covers are scalloped, each point forming a short spine. The rear gill cover tapers back into a single larger spine.

The number of preopercular spines is also one of the features used to differentiate the five different species of ruffe found across northern Europe and Asia.

Ruffe are recorded as living up to 10 years old, though most will be eaten or die long before this.

Although female ruffe can live for up to 10 years, males reach a maximum of 7 years.

Three members of the perch family are found in British waters: the ruffe, the perch and the (introduced) zander.

Ruffe are not easily confused with zander (*Sander lucioperca*), which have an elongate, pike-like body shape (hence their alternative name of pikeperch).

However, ruffe may be confused with small perch (*Perca fluviatilis*), though perch tend to grow far larger. A key distinguishing feature is that the two dorsal fins of the ruffe are fused together and therefore distinctly connected, whereas those of perch are clearly separated by a notch.

In ruffe, the preoperculum is also scalloped and spinous, unlike that of the perch.

Whilst colour is a notoriously unreliable identification feature due to variability under different environmental conditions, ruffe lack the distinct stripes on the flank and the dark spot on the rear of the first dorsal fin observed in perch.

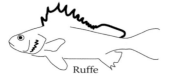

Ruffe

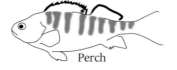

Perch

VARIABLE BODY FORMS

Ruffe tend to vary somewhat in body shape throughout their range. A 1998 study recognised four meristic (countable features such as number of scales along the lateral line or spines and rays in the fins) and morphological (shape or structure) characters in different North American and Eurasian ruffe populations.[1]

Variable features across this geographical range, some of them quite significant, include the number of soft rays in the dorsal fin, the number of preopercular spines, the length of the anal fin and the length of the caudal peduncle (the 'wrist' joining the body to the tail).

Variability in the depth of the body between different populations was also observed.

HABITAT PREFERENCES AND HABITS

Ruffe occur in a wide variety of river, lake and pond, canal, marina and also some estuarine habitats.

Ruffe are a bottom-living fish. They are more commonly found in nutrient-rich (eutrophic) waters, with a preference for still or slow-flowing water with a soft bottom and without vegetation. They can even withstand a limited degree of pollution. Though ruffe may be found in some piedmont (upper) reaches of rivers, they are absent from swiftly flowing rivers.

Ruffe often prosper in the enriched and generally murky waters of canals. Though generally inhabiting the bed and edges of the waterbodies in which they occur, ruffe can occur across a range of depths, though they appear to be more abundant in deep waters where they often co-exist with perch.

Ruffe can also occur in estuaries, most commonly in those of larger rivers, in salinities of up to 10–12% (up to one-third the salinity of full sea water). They are also found

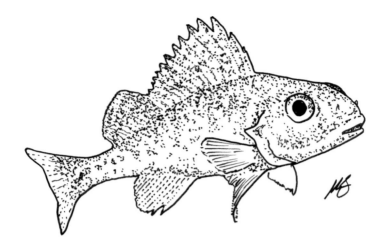

in brackish lakes across their wider Eurasian distribution. They are common in lower salinity zones of the Baltic Sea region.

Though ruffe are by no stretch of the imagination rare fish, they are only locally common and hence rare or absent outside of restricted localities. Where they prosper, they can be very abundant, and it is not uncommon to find dense, stunted populations in some waters. Ruffe are active during the day, relying largely upon sight to hunt their prey. At night, they tend to lay close to the bottom, though they may still hunt by sensing vibrations.

Ruffe are social by habit. They are generally found in shoals. These are often small but sometimes may be quite dense. As Richard Franck wrote in his 1694 Northern Memoirs:

> Ruff for the most part move all in a body. One would think them
> mutineers, because all of a piece; for if you hang but one, all
> the rest are in danger. Nor will they revolt, or retreat from their
> diet, since every one resolves to eat till he die.

H. Cholmondeley-Pennell, in his 1863 *The Angler-Naturalist*, ascribed emotional qualities to ruffe. In practice, his observations are most likely simply to be a consequence of depriving this shoaling species of the company of its kin:

> *It has been remarked elsewhere that fish are capable, under certain circumstances, of exhibiting considerable attachment for others; and this is often seen in the case of fish kept for some time together in the same receptacle. When thus situated they not unfrequently contract a strong affection for one another, insomuch that, if by any chance they are separated, they mope, or refuse their food, and in some cases actually die of grief. An instance of this attachment in the Ruffe species is mentioned in the 'Philosophical Transactions,' vol. ix.: – 'Two Ruffs were placed by Mr Anderson in a jar of water about Christmas, and in April he gave one of them away. The fish that remained was so affected that it would eat nothing for three weeks, until, fearing that it might pine to death, he sent it to the gentleman on whom he had bestowed its companion. On rejoining the other it ate immediately, and very soon recovered its former briskness.' This, if the fish were of different sexes, may possibly have been the 'pining away for love, and wasting lean,' alluded to by Burton in his 'Anatomy of Melancholy;' but if of sexes similar, then clearly it was the passion of friendship in its most platonic phase.*

THE SENSES OF THE RUFFE

Ruffe have a relatively large eye and are known to feed by sight. Moving food items are particularly attractive to them.

Ruffe also have a strongly developed lateral line extending the length of the body. The lateral line system comprises a line of sensory organs along the flanks used to detect movement, vibration and pressure gradients in the surrounding water.

The head canals of the ruffe have membranous external walls, also providing high directional sensitivity to water movement. This combination of vibrational senses enables ruffe to feed in darkness, though they are principally daytime feeders.

THE DIET OF THE RUFFE

Ruffe are a bottom-dwelling and largely bottom-feeding species, though they will pursue live food for a short distance above the bed.

The diet is entirely carnivorous. Young ruffe subsist mainly on planktonic crustaceans for the first summer of their lives, before transferring to the adult diet. The ruffe's diet at all ages is broadly similar to that of smaller perch, with which ruffe appear to be direct competitors.

The diet of adult ruffe is broad but entirely based on small animals, including both invertebrates and small fish and the eggs of larger fish. Zooplankton (small animals free-swimming or suspended in the water column) can continue to form part of the diet throughout life.

Larger invertebrate prey is eagerly hunted and consumed on an opportunistic basis. This includes various types of earthworms that happen to fall into the water or are washed in during flood events.

Bloodworms, the red larvae of chironomid midges inhabiting river and lake beds, are a favoured food. So too are midge pupae. The larvae of a range of other aquatic insects, such as caddis (Trichoptera), alderflies (Sialidae), beetles (Coleoptera) and dragonflies (Odonata), are also eagerly

ingested by ruffe. So too are smaller adult aquatic insects such as lesser water boatmen.

The ruffe's diet may also include any of a range of small crustaceans, including copepods, ostracods, freshwater shrimps and water-slaters (hog lice). They also feed on oligochaetes (varieties of worm), as well as molluscs, including snails and pea-mussels. Ruffe also feed opportunistically on small amphibians, particularly the tadpoles of newts, frogs and toads.

The ruffe's appetite for fish eggs and fry is a trait that can, as we will see, be problematic for native species when ruffe are introduced into new waters.

REPRODUCTION AND LIFE CYCLE

Ruffe usually mature in two to three years at a temperature of 10–15°C. However, in warmer waters, some individuals may reproduce after the first year of life. Ruffe head into shallower, marginal water to spawn during the spring. This generally occurs from as early as March through to approximately the end of May, depending on latitude and water temperature. They spawn in shoals on a variety of substrates at depths less than 3 metres.

Little or no sexual differentiation is observed in ruffe. Females are slightly larger than males and tend to live a little longer. Günther Sterba noted in his 1962 book *Freshwater Fishes of the World*:

> The ♀♀ may be recognised during the breeding season by their greater girth.

Female ruffe release strands of sticky eggs, male fish releasing their milt at the same time to fertilise them. These strips of eggs, white or yellow in colour, become adhesive on contact with water, leading them to stick to submerged stones, plants or woody material such as dead branches, underwater roots and overhanging tree branches dipping into the water. Individual egg diameter ranges from 0.34 to 1.3 millimetres.

Female ruffe tend to lay their eggs in two or more phases during the summer, usually separated by about 30 days. The first portion of eggs is larger than the second portion.

Ruffe are highly prolific, female fish potentially laying between 130,000 and 200,000 eggs annually. In his 1958 book *Small Fry and Bait Fish: How to Catch Them*, Kenneth Mansfield notes:

> For so small a fish the ruffe produces an inordinate number of eggs – 150,000 to 220,000 have been mentioned – but many are consumed by fish and insect larvae.

After the eggs are released, ruffe exhibit no parental protection. As a consequence, the eggs are subject to heavy predation by a diversity of fishes and invertebrates.

Surviving fertilised eggs hatch after between 5 and 30 days, depending on the temperature of the water. When first hatched, the larvae are tiny, between 3.5 and 4.4 millimetres long. They are also incompletely developed and remain immobile for 3–7 days whilst consuming their attached yolk sac and the oil globule within it. After the yolk is consumed, typically a week after hatching, the larvae become pelagic (starting to swim in open water) and feed actively on planktonic invertebrates for few days.

Alwyne Wheeler wrote in his 1969 book *The Fishes of the British Isles and North-West Europe*:

> *The young ruffe eats planktonic crustaceans for the first summer of its life; Thereafter its diet consists mainly of bottom-living animals.*

After these initial few days of independent feeding, metamorphosing juveniles switch to a benthic life (on the lake or river bed). Juveniles are initially secretive and solitary, tending to avoid others of their kind. However, they become social later in life.

Juvenile ruffe thereafter feed on a range of small invertebrates, with progressively larger animal prey items taken as the fish grow.

Survival of the larvae is poor below 10°C and above 20°C.

Ruffe are relatively short-lived, reaching five to six years or, more rarely, seven years. As noted previously, ruffe have been recorded as living up to 10 years old, females living longer than males, though this age is exceptional and does not occur in British waters.

THE NATURAL DISTRIBUTION OF THE RUFFE

Eurasian ruffe are native to most European countries, including eastern England, throughout France, except in the west, Germany, the Netherlands and Sweden. They are naturally absent from Spain, Portugal, Norway, northern Finland, Ireland, Scotland, Croatia, Serbia and Montenegro. However, they have been introduced more widely, including into parts of western France, northern Italy and Greece.

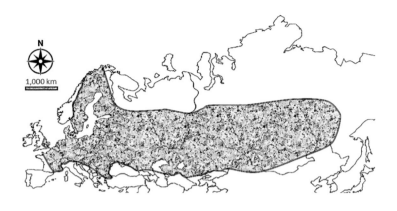

The natural range of ruffe in the British Isles reflects our geological history. Until the last Ice Age (between 6,500 and 6,200 BC), Britain was connected to continental Europe by 'Doggerland', a land mass that now lies beneath the southern North Sea where it is better known as the Dogger Bank.

During this time, the Rivers Thames, Rhine and Scheldt formed the Channel River – today the English Channel – carrying their combined flow to the Atlantic. Ruffe occurred across the Channel River catchment, which included present-day British catchments from the Humber southwards to the Thames. Beyond this natural range, terrestrial barriers between catchments prevented ruffe and many other species of fish from spreading westward and northward across the land mass.

However, towards the end of the Ice Age, a gigantic ice lake to the north of Doggerland broke to release a megatsunami. This catastrophic event cut through and inundated the former land bridge, separating the contemporary British Isles from continental Europe. The consequent limited natural British distribution of ruffe (as indeed many other fish species such as barbel, gudgeon and burbot) across river catchments of Eastern England formerly connected with the Channel River is explained by this geological history.

It is due to this same geological history that ruffe are also naturally absent from Ireland. *Giraldus Cambrensis* ('Gerald of Wales' [c.]1146–[c.]1223), a medieval clergyman and chronicler of his times, published an account of *The History and Topography of Ireland*. In this book, he recorded that "... *pike, perch, roach, gardon, gudgeon, minnow, loach, bullheads and verones* ..." were absent from Ireland. Giraldus observed that all the Irish species of freshwater fish known to him could live in salt water as migratory or brackish-tolerant species. Of these, he listed brown trout, Atlantic salmon and arctic charr as well as pollan, three-spined sticklebacks, European eels, smelt, shad, three species of lamprey, and the increasingly rare common sturgeon.

Although ruffe are found in some larger estuaries in continental Europe, particularly around the Baltic Sea, their salt tolerance is limited to water only at a third of the salinity of sea water. Consequently, they were not included amongst the fully salt-tolerant fishes able to colonise Ireland's freshwater systems without man's interference. Ruffe have not, at least to date, been subsequently introduced to Ireland.

In Asia, ruffe are native to the rivers, lakes and brackish sea coastal waters of most of the former USSR. This range extends nearly as far north as the coast of the Arctic, Baltic and North Seas and, to the south, to the Aral, Caspian and Black Seas. Ruffe occur widely throughout Siberia with the exception of the Amur River, Lake Baikal and Transcaucasia.

PREDATORS OF RUFFE

As a small fish, ruffe are prey for a wide range of predatory fishes such as zander, pike and perch. Although their spikey front dorsal fin and gill covers are believed to dissuade at least some predation, the fish are sufficiently small to be readily swallowed by larger predators.

Other aquatic predators likely to eat ruffe include birds such as herons and egrets, and smaller individuals may be taken by kingfishers.

A study found that ruffe had first appeared in Loch Lomond in 1982, subsequently increasing exponentially in population over the following decade. Loch Lomond's ruffe were found to have become the primary prey species for cormorants (*Phalacrocorax carbo*), herons (*Ardea cinerea*) and northern pike (*Esox lucius*).[2]

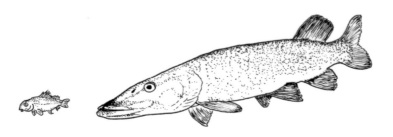

Mammalian predators of ruffe include otters, the predominant diet of which comprises many more small benthic species than some anglers appreciate. This is due to their essentially 'truffling' feeding habit on the bed and margins of waterbodies.

RUFFE INTRODUCTIONS IN EUROPE

Introductions of species beyond the native range within which they co-evolved is always potentially problematic. Invasive introduced species in British waters include aquatic and riparian plants such as Australian swamp stonecrop (*Crassula helmsii*) and Himalayan balsam (*Impatiens glandulifera*), various invertebrates including American signal crayfish (*Pacifastacus leniusculus*) and 'killer shrimp' (*Dikerogammarus villosus*) as well as non-native fishes including topmouth gudgeon (*Pseudorasbora parva*).

As noted previously, ruffe have been introduced widely beyond their native range, often with adverse consequences. Some of the attributes of aquatic organisms that predispose them to become invasive are that they are abundant and widely distributed in their original range, with wide environmental tolerance, short generation time with rapid growth and early sexual maturity. Characteristics facilitating inadvertent or deliberate transport through human activities can also expedite their spread, including carriage in ship ballast water or trade for ornamental, recreational, gastronomic or other purposes.

A 2016 study found that ruffe were flexible with regard to chemical, physical, biological and habitat requirements across their different life stages, adapting them to a range of environments.[3] Together with their broad diet, rapid maturation, relatively long life and habit of batch spawning, these features enable ruffe to successfully colonise new habitats following introduction. As they feed on a variety of eggs and small fish and invertebrate prey, rapidly increasing populations of ruffe in new waters can suppress populations of native fish species, giving cause for concern.

Problems can result even within a country or region where a species is naturally present if introduced beyond its native range. Where artificially introduced beyond their natural eastern English range, ruffe have been found to disrupt some of the ecosystems into which they have become established with unpredictable and potentially catastrophic, or at least harmful, effects.

Ruffe are not native to northern England. However, they have been present in the English Lake District for decades and have caused significant ecological harm in some lakes. They have been recorded at Bassenthwaite Lake, Derwent Water and Thirlmere over many years and in Rydal Water more recently. Scientists made the first conclusive sighting of ruffe in Lake Windermere in 2019, although the presence of the fish had been inferred by an environmental DNA (eDNA) survey in 2015. (eDNA surveys sample water for the presence of DNA [genetic material] released into the water by species that are present.) Some of the impacts of these introductions are discussed later in this book.

Transfer as live bait by pike anglers is commonly assumed to be the cause of these introductions. However, as ruffe are small fish, this may not necessarily be the case. In fact, Kenneth Mansfield wrote of the ruffe in his delightful 1958 book *Small Fry and Bait Fish: How to Catch Them*:

> As far as I know ruffe are never used for live-bait except when nothing else is available.

An undated article by the Canal and River Trust[4] endorses this view:

> Whilst I would not pretend to have much pike fishing knowledge, I have yet to come across a single pike angler who has ever claimed to have used ruffe as a livebait.

Introduction of ruffe into these deep, cool lakes has had devastating impacts on native fish species adapted to these glacial habitats. These include fish considered 'glacial relics', left behind in cold water systems as glaciers retreated explaining their often-scattered distribution in isolated deep-water lakes from north Wales to northern England and Scotland. These fishes include European whitefish (*Coregonus lavaretus*), vendace (*Coregonus albula*) and Arctic charr (*Salvelinus alpinus*), which require good-quality, well-oxygenated water in deep lakes.

Threats are posed by contemporary land use as well as the discharge of effluent contributing to various forms of pollution that can deplete oxygen in deeper layers of these lakes, as well as the blinding of open spawning gravels with silt. The introduction of coarse fish species beyond their native ranges, including roach and ruffe, can also compromise the requirement of native species for spawning habitats free from excessive egg predation.

Ruffe, in particular, have been implicated as a major factor in the extinction of vendace from Bassenthwaite Lake in the English Lake District. In Wales, they pose similar threats to the Welsh population of Gwyniad (the local name for the European Whitefish) in Llyn Tegid (Bala Lake).

Though formerly absent from Scotland, ruffe were first discovered in Loch Lomond in 1982.[5] Between 1982 and 1992, ruffe became established and the population grew exponentially before stabilising at a high level. The ruffe also expanded their range throughout the loch and beyond it into the slow-flowing influent and effluent tributaries.

The ruffe in Loch Lomond were found to prey mainly on benthic macroinvertebrates (large invertebrates living on the like bed) but also the juveniles and eggs of native fishes. The most significant amongst these vulnerable native fish species is the powan (the local name of the European whitefish, *Coregonus lavaretus*), a fish of high national nature conservation value. Research found little evidence of overlap in the diets of adult ruffe, perch (*Perca fluviatilis*) and brown trout (*Salmo trutta*).

Ruffe are now abundant throughout Loch Lomond, raising concern about their impacts on the endemic powan due to predation on their eggs. An undated Canal and River Trust article notes:

The ruffe is despised by most Scottish fish biologists for its introduction into Loch Lomond. This seems to have happened around 1982 and has impacted significantly on the ecology of the Loch Lomond fish community. Ruffe are one of six coarse fish species to have been illegally introduced into Lomond over the years, the others being chub, bream, crucian carp, dace and gudgeon. The ruffe is guilty of being a prolific consumer of the eggs of the resident whitefish, known locally as the powan. Ruffe are potentially endangering the long term survival of this species.

Ruffe are still spreading westwards from their former eastern native range in England through human interventions. For example, in the Kennet Avon and Canal, linking the River Kennet (a major tributary of the Thames catchment) with the predominantly south-western flowing Bristol Avon catchment that discharges to the Severn Estuary, ruffe were accidentally introduced with transfers of stocks of larger fish species.

Ruffe have not only become established in the Kennet and Avon Canal but also proliferated at least to date in the western end of the canal. In July 2017, I was the first person to catch and confirm two ruffe specimens from the Bristol Avon river itself, this from the Feeder Canal to the floating harbour in Bristol. As the Feeder Canal is 18 kilometres downstream from where the Kennet and Avon Canal joins the Bristol Avon River at Bath, ruffe are likely to have become well-established in the lower Bristol Avon River.

Elsewhere in Europe, ruffe have become established in Lake Geneva (Switzerland and France), Lake Constance (the Bodensee bordering Austria, Germany and Switzerland), Lake Mildevatn (Norway), the Camargue region of southern France and northern Italy. They are inherently problematic in all of these regions.

RUFFE PARASITES

One of the additional dangers of transfer of ruffe and other fishes, both within their natural ranges and beyond, is that they can also carry parasites with them. These parasites in turn can then infest other fishes, potentially becoming established in new ecosystems with unpredictable consequences.

Various studies have analysed the parasites carried by ruffe. The range of parasitic organisms infesting ruffe is diverse. Amongst them are cestodes (tapeworms), other internal digenean and external monogenean flukes, as well as nematode worms. Parasite loads tend to increase in polluted water.

It has been reported, for example, that the mongenean fluke *Dactylogyrus amphibothrium* was transferred to the United States with introduced ruffe.[6]

RUFFE INTRODUCTIONS IN AMERICA

Ruffe have been introduced inadvertently into the Great Lakes region of North America. They were first reported in Lake Superior in 1987. They have since been found more widely in Lake Superior as well as in Lake Huron and Lake Michigan bordering the US states of Minnesota, Wisconsin and Michigan and the Canadian province of Ontario. These ruffe were reputed to have been introduced in ballast water from ships. (This is a more likely transfer mechanism than theories of spread as unwanted live baits.)

The introduction and spread of ruffe in North America as a non-native fish species is taken very seriously. Ruffe have been the subject of significant levels of research by fisheries scientists seeking means to eradicate them from the Great Lakes network due to the damage they are doing to native fish species.

Ruffe have been found to negatively impact native sportfish populations in the Great Lakes system, including yellow perch (*Perca flavescens*). Impacts derive directly from competition for food and habitat, and through heavy predation of native sportfish eggs. Their rapid reproductive and growth rates mean that ruffe can rapidly become the most dominant fish in local areas, putting pressure on native species and contributing to their decline.

Efforts are being invested to halt the potential progressive spread to each of the Great Lakes and many connected inland waters before greater damage can ensue. Ruffe have already become the most numerous fish species in the St. Louis River watershed, which covers 3,584 square miles (9,283 square kilometres) located at the head of the Great Lakes.

One of the primary methods developed to control ruffe entails increasing the population of their predators, including walleye (*Sander vitreus*) and northern pike (*Esox lucius*). Other methods considered include chemical controls, involving both poisoning large schools of ruffe and the use of alarm pheromones (hormones released externally to the body) to repel the fish from spawning sites.

MORE INFORMATION ABOUT RUFFE

Ruffe, along with the two freshwater species of stickleback and both loach species, bullheads, minnows and several smaller fish species found in Britain, are often dismissed as 'minor species'. As all fish play integral roles in ecosystems, and are fascinating in their own right, this feels wrongly dismissive to me.

However, ruffe do not feature in many books on coarse fishes. These range from Arthur P. Bell's 1926 *Fresh-water Fishing for the Beginner*, Eric Marshall-Hardy's 1943 *Coarse Fish*, and Nick Giles' 1994 *Freshwater Fish of the British Isles: A Guide for Anglers and Naturalists*.

If you want to know more details about the biology of ruffe and other British freshwater fishes, I can point you towards three of my other books that do include them:

- *The Complex Lives of British Freshwater Fishes* (2020). CRC/Taylor and Francis, Boca Raton and London.
- Everard, M. (2013). *Britain's Freshwater Fishes*. Princeton University Press/WildGUIDES, Woodstock, Oxfordshire.
- Everard, M. (2008). *The Little Book of Little Fishes*. Medlar Press, Ellesmere.

NOTES

1 Stepien *et al.* (1998)
2 Adams and Maitland (1998)
3 Gutsch and Hoffman (2016)
4 Canal and River Trust (Undated)
5 Adams and Maitland (1998)
6 Cone *et al.* (1994)

Three

In their humorous 1925 work Fishing: Its Cause, Treatment and Cure, H.T. Sheringham and G.E. Studdy comment upon the perpetual hunger of ruffe shoals:

> *Annoys the perch-fisher more than the bleak annoys the*
> *roach-fisher. Pope are chiefly remarkable for an appetite which*
> *cannot be appeased and for never growing any bigger. They*
> *could not be any smaller.*

Singularly dismissive of the virtues of ruffe fishing, or at least the catching of ruffe whilst angling for other fishes, Henry Coxon in his 1896 book *A Modern Treatise on Practical Coarse Fish Angling* then proceeds to commend their flavour. (The 'old writer' in question is none other than Izaak Walton.)

> *With regard to the ruffe, or pope, I regard him as a 'fish driver.'*
> *By this I mean the family take possession of a secluded spot*
> *or corner and their presence seem to drive all other fish away.*
> *Often enough, when fishing with a tight line, for roach, I have*
> *killed ruffe after ruffe and have left the spot in disgust. The*
> *dorsal fin of the fish is single, and its flesh is excellent. Says*
> *an old writer: 'No fish that swims is of a pleasanter taste; he*
> *is a greedy biter; and they will usually lie, abundance of them,*
> *together, in one reserved place, where the water is deep, and*
> *runs quietly. Just so, and there they may lie so far as I am*
> *concerned.*

DOI: 10.1201/9781003311034-3

As described by Alwyne Wheeler in his 1969 book *The Fishes of the British Isles and North-West Europe:*

> *The ruffe is too small to possess any interest to anglers; indeed its bold biting at bait intended for better fish is more often a nuisance than anything.*

The ruffe is indeed a bold biter on worm, maggot or other live baits, even in chilly weather, and is often a welcome friend to the match angler fishing in tough conditions on a canal. In fact, many of us welcome a ruffe in a mixed bag or else target them specifically.

RUFFE LOCATION

Mr Lane provides a few notes on the general appearance and angling methods for the ruffe in his 1843 *The Diary of A.J. Lane: With a Description of Those Fishes to be found in British Fresh Waters,* adding:

> *This fish is very common in the Thames and many other rivers with still places & which are rather sluggish. Prefers gravelly places and is gregarious and voracious. They are seldom caught above six inches long.*

Ruffe prefer still or sluggish water and are often more highly concentrated around features in the water. They are fish of the bed and banks of the waterbody – a bait up in the water is rarely or never taken by these fish – so are most effectively fished for on the bottom or against any structures. They also thrive in murky water. Canals seem to be a favoured habitat where, if present, ruffe can be prolific if not necessarily individually large.

Vertical structures, in particular, are attractive to ruffe, providing both cover and a larder of animal matter including invertebrates, small fish and eggs. Ruffe tend to graze or loaf around vertical canal banks or bank reinforcement of any kind, as well as around submerged posts and branches, the shelter of moored boats and lock gates.

The diligent angler can have great sport working the margins of a canal, river or lake until they find a shoal of these generally obliging fish.

RUFFE BAITS

Ruffe are dedicated carnivores. They feed opportunistically on a wide variety of invertebrates, small fish and the fry of larger fishes as well as their eggs.

Consequently, readily sourced and favourable baits for ruffe include maggots and worms. The maggots can be of any colour, though I and many other anglers find that red maggots work best. This may be related to bloodworms forming such an important and widespread food item favoured by ruffe and many other fish species.

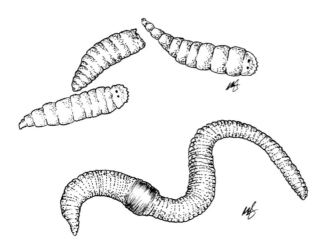

A variety of types of worm also work, including small worms or chopped up pieces of larger worms. In terms of small worms, dendrobaena are favoured. Brandlings also can work if necessary, although, as they exude a pungent yellowish liquid as a defence mechanism, they tend to be less palatable.

Walton's Piscator has a few words to say about preferred ruffe baits in *The Compleat Angler*:

You must fish for him with a small red worm; and if you bait the ground with earth, it is excellent.

Other small invertebrate baits can work well if they are readily obtained. Caddis larvae, freshwater shrimps and almost any other aquatic invertebrate found where you intend to fish will also be located by ruffe. Tucking a pond net into your fishing bag can yield you suitable bait. Potentially, fish fry are also suitable bait, though hooking them tends to kill them, rendering the fry far less effective.

A friend of mine who used to be a keen and expert match angler swore by bloodworms and joker (the smaller larvae of smaller midge species) for targeting ruffe. Bloodworms are probably the most prolific source of natural food for bottom-feeding fishes, such as gudgeon, bream, carp, tench as well as ruffe. The bright red colour giving these larval insects their common name is due to high levels of pigments within the body, scavenging oxygen from their surroundings and so enabling them to thrive in low oxygen conditions.

Fine wire hooks are necessary down to sizes 22 or 24 for bloodworms or clumps of joker, or even 26 for single joker. These present the bait just off the bottom using fine lines and under the finest of pole floats, ideally wire-stemmed with a fine bristle tip, to indicate the tiniest of bites. Bloodworm and joker are particularly effective in the winter months when bites are hard to come by, their small size and active movement making them highly attractive to a range of fish species.

Though a favourite with some match anglers, there is some controversy about the use of bloodworms and joker. In fact, in some areas of the country, these baits are commonly banned in matches as they are perceived to offer an unfair advantage to anglers able to afford them, where the supply of this bait is limited, but also due to the potential damage that can occur when the tiniest of fish are hooked. No-one, however, doubts their effectiveness, and bloodworm and joker are much more commonly used in mainland Europe.

Bloodworm and joker can also be mixed with other baits to great effect to attract and catch ruffe. Balls of joker introduced into relatively still swims in canals, cuts and marinas can be a powerful attractor, with larger baits such as red maggots offered on the hook. Using this method, or with bloodworm or joker on the hook, movement is a key attractor.

Lively hook bait presented slightly off the bottom, further animated by periodically raising and lowering the float or dragging it slightly to create movement, is a highly effective means to induce a take from a ruffe.

In fact, bloodworm fishing has a long history in Britain. As Francis Francis wrote in 1867 in *A Book on Angling*:

We northern anglers generally term our groundbaits as 'feeds' this we speak of bloodworm feed, yellow feed (squatts) and bread feed. In my early days, the most successful ground bait was bloodworm feed. In almost all northern matches today the use of bloodworm is barred but wherever they are allowed and an expert bloodworm fisher is a competitor, you may confidently expect to see his name in the prize list.

On matters of bloodworm and joker fishing, I defer to the greater expertise of various friends and the writings of match anglers; in fact, I can barely see let alone tie or bait a size 24 or 26 hook!

However, worms and maggots – pinkies, maggots or squats – are readily and cheaply obtained and are also tough and remain active. They are therefore generally the most convenient and effective baits for ruffe.

A common theme with all these types of bait, as well as lures, is that the colour red seems to be highly effective. As observed, this may be due to bloodworms forming such an important natural constituent of the diet of ruffe and that of a range of other fish species.

In my experience, and that of other ruffe hunters that I know, movement matters. Meaty baits such as small cubes of luncheon meat, attractive to many species of freshwater fish ranging from roach and crucian carp to bream and tench, seem to elicit no interest from ruffe if they are static. Live and meaty baits are, however, greeted with gusto as are lures. Much more on lures is to follow later in this chapter.

During the drafting of this book, I was told by one angler that he had once caught a ruffe on a grain of corn. As it turned out, this was when the bait was 'on the drop'. I will be far from alone in having predatory fish including pike, trout and perch taking bread, luncheon meat, mini-boilies as well as corn when retrieving tackle, the moving bait imitating live prey or at least eliciting a predatory response. Movement, it seems, is a compelling part of the lure of any bait offered to attract ruffe.

PRESENTATION FOR RUFFE

The fishing tackle trade is smart at spotting niches to part us gullible anglers from our heard-earned cash! Aside from the plethora of novel baits, rigs and pimped reels and bite indicators, float rods are sold with either hollow or spliced tips, or in waggler or stick float forms, even though these differences in a well-designed rod are largely to catch the angler rather than more fish. Ready-made rigs suited to different species and sizes of fish are also widely available through retail outlets. Float and quivertip rods of subtly different actions and lengths best suiting them for often highly specific purposes are also offered. The market is as diverse and large as the tastes and the depth of the pockets of the customer base that it serves.

The days of the generalist rod that I grew up with – a wooden or fibreglass float rod that doubled up for legering with or without a screw-in quivertip or swingtip – is long gone despite being generally entirely adequate for a host of purposes. So why then have we not seen a ruffe specialist rod hit the tackle shop shelves? Why no ruffe-specific bolt rigs, floats or other widgets? Is the tackle trade missing a massive untapped market?

The reality is that ruffe are neither fussy eaters nor generally all that hard to catch. Nor are they a prestigious target species for the specialist, specimen or, in all but the hardest conditions, match angler. It is only people like me and, presumably, you also as a reader of this book who every now and again fancy setting out our stall to catch a few ruffe for the sheer fun of it, generally doing so in good company! And, when we do, the mahseer, tench and salmon rods are left at home as we venture out onto the bank with the most rudimentary rod and reel set-ups to present our maggots or scraps of worm under a float or leger, or even on a freeline.

Fishing with light line and balanced light tackle adds to both sensitivity and enjoyment, with hooks ranging from 20 to 14 depending on bait size. But the reality is that, for most of the time, nothing more sophisticated is necessary.

The only specifics of presentation for ruffe are that these fish are present in the water you are fishing and that a 'meaty' and ideally live bait is presented on or near the bed or edges of the river, lake or canal to intercept these predominantly bottom-dwelling fish. A bait presented any significant distance above the bed is likely to be ignored by even the hungriest ruffe.

Beyond this constraint, tackle choice need not be specialised. As Kenneth Mansfield put it in his 1958 book *Small Fry and Bait Fish: How to Catch Them*:

> The lightest of tackle can be used, with a small float. The bait should fish on or very close to the bottom. It is advisable to strike as soon as the float bobs in order to lip-hook the fish: if a ruffe is given a few seconds spare it will gorge the bait.
>
> Most of the ruffe caught are taken by anglers seeking more important quarry and they can undoubtedly be a nuisance, especially in waters where they subsist in thousands, as in some parts of the Broads.

This advice about rapid striking is well offered as ruffe do tend to gorge the bait rapidly, as do small perch. Deep hooking is no fun for either the angler or the ruffe! Every angler should carry a disgorger, a cheap tool that will help matters if the bait is taken deeply. At this point, I will repeat the best piece of angling advice ever given to me. Always carry three disgorgers: one to use, one as a spare and a third to give to any other angler you meet who does not have one!

Light leger tactics too can serve the ruffe angler well. Again, nothing fancy is needed. A light link leger or paternoster to present worm, maggot of other baits on or near the bed is sufficient, but again heeding the advice to strike quickly to avoid deep hooking. As Izaak Walton put it in *The Compleat Angler* regarding ruffe fishing:

> . . . and he is also excellent to enter a young Angler; for he is a greedy biter, and they will usually lie abundance of them together in one reserved place where the water is deep, and runs quietly, and an easie Angler, if he has found where they lie, may catch forty or fifty, or sometimes twice so many at a standing.

LURE FISHING FOR RUFFE

We have a narrow view of which species of fish are predatory and which are not, and consequently which can be caught on lures. In fact, virtually all species of fish are, to one extent or another, predatory. The juvenile life stages of virtually all British freshwater fish species rely on small invertebrate food. Very many remain opportunistic throughout their entire lives. This includes those not normally considered predatory, such as roach, which will readily feed on a variety of available animal life including invertebrates but also fish fry and eggs. Consequently, all of these fishes can be caught on appropriately small lures presented where the fish can be expected to be feeding. This is particularly so in warmer water, as higher temperatures increase the metabolic rate of the fish, heightening their predatory instincts.

Ruffe are, of course, entirely predatory. Consequently, they are amenable to being fished for with lures of suitably small size, imitating their various invertebrate and small fish prey. This is an exciting new horizon of fishing for ruffe and other fishes. It is an approach that is beginning to become increasingly popular in Britain.

As we have seen, ruffe are particularly attracted to live food, sensing it both by sight and by vibrations in the water. As Kenneth Mansfield notes of ruffe in his 1958 book *Small Fry and Bait Fish: How to Catch Them*:

> *They appear to find their food more by sight than smell and are daylight feeders.*

If you want a fully authoritative and contemporary guide to lure fishing in all its great diversity including, as they put it, fishing for the 'mini-monsters', see the great 2019 book *Hooked on Lure Fishing* by Dominic Garnett and Andy Mytton. There is not the space here to go into depth about specialist rods, reels, grips and more; we'll just look at specifics as they relate to ruffe.

Richard Widdowson, a specialist in this type of fishing, recommends red, white and pink soft plastic lures for ruffe. Worm-type soft plastic lures are recommended, threaded onto weighted jig heads. However, a huge variety of micro-scale soft plastic lures is available imitating small fish, worms, crayfish, grubs, dragonfly larvae and various other insects and organisms, and in a bewildering range of colours and sizes.

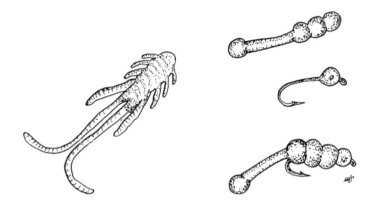

Soft plastic lures of a size suitable for mimicking bloodworms are ideal for targeting ruffe. These can be up to an inch (two-and-a-half centimetres) in length, though up to two inches (five centimetres) where you expect big ruffe.

These small plastic lures are threaded onto weighed jig heads from size 10 down to 16, at weights of 0.5 or 1 gramme. An alternative is to attach a shot in front of a hook onto which the soft plastic lure is threaded. These micro-lures are also likely to be attractive to a wide range of other types of freshwater and marine fish.

Fine lures are best tied to a length of fluorocarbon line of 5–6 lb breaking strain, mated at the end of light braid reel line. Fluorocarbon line has a key advantage over nylon as it becomes all but invisible underwater. The fluorocarbon trace line at this breaking strain balances fine presentation with sufficient strength to help extract the lure from sunken branches and other snags, as well as dealing with the occasional big pike as even quite big fish are far from averse from intercepting small lures.

Braid reel line is recommended as it has zero stretch and so transmits the activities of the lure and taking fish more directly to the angler holding a suitably delicate and short rod, helping them feel for often subtle bites.

The lure is bounced along the bed or the side walls of the river, lake or canal. This is where ruffe may be hunting for small fish, including sticklebacks, the fry of larger fish and a range of insects and other invertebrates. But gudgeon, roach, bream and many more species of fish, as well as often larger predators such as pike and perch, may also intercept the lure.

As the angler's connection with the lure is so direct, instant striking generally avoids problems of deep hooking, and even occasional pike are generally hooked in the 'scissors' avoiding bite-offs.

Ruffe, and many of these species, are generally to be found in proximity to submerged features such as nooks in bank reinforcements, sunken branches, weed beds and also shopping trollies. Fish are often encountered just off the bottom, so bouncing the lure across the bed can be effective in locating them. However, fishing vertically along banks and around submerged obstructions can be highly effective. The micro-lure is often fished almost vertically, with the angler keeping their silhouette away from the water's edge as they search out likely submerged features.

Some experts of this type of fishing also have success dipping their lures in flavourings, adding extra attraction to the primary stimulus of movement of the lure. Scent can apparently make a difference in the dirtiest of water, worm additives reportedly working particularly well for ruffe.

An alternative approach to presentation of fine soft plastic lures is drop-shotting. Drop-shotting entails threading a soft plastic lure onto a hook connected to a line such that it sticks out horizontally. The angler gently jiggles the rod to make the lure vibrate in the water at a fixed depth, set between a weight at the end of the line and the reel line. Minimal vibrations of the lure are imparted by subtle movements of the rod tip. Typically, a palomar knot is used to make the hook stand out at 90 degrees from the almost vertical line.

Generally, the terminal weights used in drop-shotting have a spring clip that grips the line, enabling easy movement and hence adjustment of the working depth of the lure. As ruffe feed mainly on or near the bed, this distance is not set very deep unless the lure is being worked against a vertical surface such as a canal wall or similar feature.

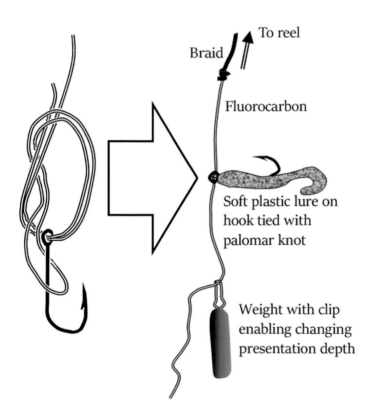

To reel

Braid

Fluorocarbon

Soft plastic lure on hook tied with palomar knot

Weight with clip enabling changing presentation depth

Although ruffe are generally thought to feed mainly by sight, remember that they are also highly sensitive to vibrations and other movements in the water. Therefore, ruffe, along with various other species of freshwater and marine fish, can also be caught well into dusk or even full night-time. This sense also means that they are also catchable when fishing murky water, such as canals that are frequently disturbed by boat traffic.

Part of the pleasure of this type of fishing is that it is light and entails minimal kit. This enables the angler to cover significant distances along river, lake or canal banks, searching out fish in likely looking niches rather than waiting in the hope of fish coming to them.

Whilst this form of light lure fishing is relatively novel in Britain, ruffe are very commonly caught on jigs by ice fishing anglers in Russia. Fine jigs are lowered through holes bored with an auger through thick ice on rivers and lakes in extreme cold conditions. These intrepid anglers are generally targeting perch, pike, whitefish, zander, arctic charr, burbot and other more desirable fish species under the ice. Ruffe are often perceived as a nuisance due to their keenness to intercept the angler's lure.

FLY FISHING FOR RUFFE

It is also possible to catch ruffe by fly fishing. As ruffe are predators active on the bed and around submerged features in still and slow-flowing waters, we are not, needless to say, talking here about dry fly fishing. Ruffe simply will not rise to a fly presented on the surface.

A key feature of any wet fly suitable for ruffe fishing is that it is heavy enough to sink rapidly, enabling it to be retrieved tripping the bed of the waterbody. Flies tied with a metallic bead head are ideal. Movement is the key trigger, possibly with a flash of red for added stimulation.

One of my favourite wet flies for all coarse fish – one on which I have caught everything from roach, dace, chub and three species of trout (brown, rainbow and brook) from British waters in addition to South African yellowfish and Himalayan mahseer – is the gold-ribbed hare's ear, or GRHE. Hook sizes 10, 12 or 14 suit most needs, the fly generally tied with a gilthead bead. The trusty old GRHE imitates pretty much any of a range of invertebrates found in flowing and standing fresh waters.

Given the need to bounce the wet fly across the potentially snaggy bed or margins of the waterbody, jig flies are a good choice as they fish with the hook point oriented upwards. A pattern I have devised uses a slotted tungsten or gilt bead at the head of a jig hook with half a rubber bloodworm as a tail and a body of dubbed red fur or peacock herl. The whole thing closely emulates bloodworms or other types of small red larvae or worms.

I enjoy fly fishing for pike and perch with big flies fished with floating lines. But, in my view, much fly fishing in deep water with weighted lines and heavy flies differs only semantically from spinning with artificial lures. (Sorry to offend any fly fishing purists, though I suspect that not many will be reading this book!)

It is not only entirely possible to fly fish for ruffe, it is also fun to do so in shallow water using five-weight or lighter floating fly lines with a sunken tippet. However, lure fishing is generally a more efficient, but otherwise not dissimilar, alternative to heavier wet fly fishing in deeper water.

RUFFE RECORDS

The British record rod-caught ruffe at the time of writing – a hostage to fortune as the record may be broken any day! – is held by a magnificent fish weighing 5 ounces 4 drams. This mighty ruffe was caught by Ronnie Jenkins on 7 August 1980 from West View Lake in Cumbria.

Ronnie Jenkins' monster ruffe displaced the former record of 5 ounces taken in 1977 by P. Barrowcliffe from the River Bure in Norfolk. Barrowcliffe had broken the previous record held by B.B. Poynor for a fish of 4 ounces landed in 1969 from the River Stour in Warwickshire. B.B. Poynor's fish had in turn surpassed the former record ruffe of 3 ounces 9 drams, caught in 1963 by A. Cartwright from the River Mease on the border of Leicestershire and Staffordshire.

It was reported recently in the angling press that Lincolnshire-based angler, Kerry Goldsmith, caught a 6 ounce ruffe but returned it after weighing without claiming the record.[1]

Delving back even further into history, Kenneth Mansfield notes in his 1958 book *Small Fry and Bait Fish: How to Catch Them*:

There is a record of a 5 oz. ruffe caught in the Domesday Deeps, near Shepperton, but there is no mention of length.

RUFFE MATCH CATCHES

Even where they are prolific, ruffe are rarely a strategic angling target for the match angler. In part, this is because profuse populations rarely comprise fish of weighty individual size. But also, ruffe generally occur where there are mixed populations of other species of fish. The smaller ruffe tend to be displaced by larger species when they come onto the feed.

Consequently, ruffe tend to feature in match fishing bags as part of a mixed bag of species, or as a minor component of other more directly targeted coarse species. In fact, match anglers tend to be somewhat downhearted when all they can attract to their baits are small ruffe, as this can indicate that there is little else in the swim.

They are therefore more of a 'make-weight' than a principal target for the match angler.

RUFFE AS BAIT

The introduction of ruffe into the English Lake District, Loch Lomond and other waters across Britain has frequently been blamed on pike anglers bringing them along with them as bait. This has not been proven.

The same is true of the suppression of fish populations of conservation and sporting importance in the English Lake District as well as in Loch Lomond, and in other locations such as Lake Bala (Lynn Tegid).

Also, as observed when considering lure fishing, the ever-hungry ruffe is sometimes seen as a nuisance by Russian ice-fishermen as they beat more desirable species to jigs presented under the ice.

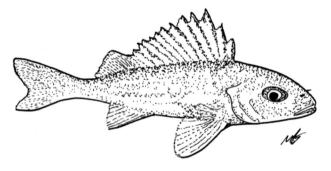

However, though easily sourced where they are abundant, and also robust, tolerant of poorer water quality and hence easy to store and transport, ruffe are quite small relative to normally sized pike baits.

As noted earlier in this book, a range of other authors and I cast doubt on assumptions that live bait transport is a major route for ruffe introductions.

Other factors may be more significant, including transfer in ballast water, considered the likely source of ruffe introductions into the American Great Lakes, or as sticky strings of eggs adhering to moist material (nets, waders, boats, etc.) moved between waterbodies.

RUFFE AS PROBLEMS

As discussed in some depth previously, ruffe have been introduced beyond their natural range not only across Europe but also into the Great Lakes region of North America, spreading further into adjacent catchments in the United States as well as Canada.

In these places, ruffe have had significant impacts on native species, with considerable nature conservation and sporting implications and associated socio-economic ramifications.

RUFFE: THE ULTIMATE SPORT FISH?

Ruffe are not weighty fish like carp, pike or wels catfish. Neither will they test the angler's tackle to its limits with their fierce runs. Nor are they, with the best will in the world, a wily adversary. However, they are seemingly always willing to take a bait or lure, and, for that, we thank them!

But what, in the real world, is a sport fish? It is a fish that gives us sport, pure and simple, a fish we enjoy seeing and catching. On that score, ruffe are fine sport fish that, on suitable tackle and in particular with novel approaches, can be a pleasure to pursue and catch.

For me, ruffe-hunting is something I most enjoy with good, like-minded friends, whilst for others it is a solitary preoccupation. There is skill in working the margins to locate the fish and in targeting bigger specimens amongst them.

The ultimate sport fish? Well, maybe not in the way that inspires breathless media headlines. But these fish, approached with suitable finesse and innovation, and with the right attitude, are most definitely a worthy angling quarry!

NOTE

1 Canal and River Trust (Undated)

. . . he is an excellent Fish; no Fish that swims is of a pleasanter taste.

Ruffe are said to be good eating, with firm tasty flesh especially when fried. Some authorities recommend ruffe soup.

Bent J. Muus and Preben Dahlstrom state in their 1967 Collins Guide to the Freshwater Fishes of Britain and Europe:

Its flesh is tasty, especially when fried, but the fish is too small to be valuable as a table fish, although ruffe soup can be recommended.

Ruffe were at one time widely exploited for food in North Germany, and the fishery in eastern-Europe was once important.

Alwyne Wheeler records in his 1969 book The Fishes of the British Isles and North-West Europe:

It is not commercially fished, except that at one time in North Germany it was exploited for food. Ruffe are said to be good eating.

Adding to this, Kenneth Mansfield writes in his 1958 book Small Fry and Bait Fish: How to Catch Them:

In north Germany the larger ruffe from brackish waters are considered a luxury and form a local dish of some repute.

In their 1967 Collins Guide to the Freshwater Fishes of Britain and Europe, Bent J. Muus and Preben Dahlstrom relate:

At one time the fishery in eastern Europe was important. In the eastern Prussian haffs the ruffe which seem sensitive to sounds, were formerly driven into stake-traps by 'clap boards' (the banging of boards which have ends sticking into the bottom).

However, there are now no remaining commercial ruffe fisheries as this species is too small to be valuable as a table fish. The main perceived interest in the ruffe in fisheries today is as a potential competitor for food with other more economically valuable sport fish or as a predator on their fry and eggs.

Many other writers are also effusive about the value and quality of ruffe as food. Reputedly relating to the eating qualities of the ruffe, the 1496 *Treatyse of Fysshynge with an Angle*, attributed to Dame Juliana Berners, notes:

> *The Ruf is a right and holsom fysshe.*

To this, J.H. Keene added, in his 1881 book *The Practical Fisherman*:

> *I much prefer the ruffe, so far as its flavour is concerned, to even the delicate sweet gudgeon or its cousin, the perch. It, however, requires careful cooking, not that it is a fish to which it is necessary to add all sorts of condiments, but because to over-fry it or over-bake it (with bay and rosemary) is to spoil a certain nutty flavour which a ruffe from the clear river at the latter end of July possesses.*

European ruffe are also not listed under the Berne (or Bern) Convention (the Convention on the Conservation of European Wildlife and Natural Habitats that came into force in 1982). However, the scarcer schraetzer or 'striped ruffe' (*Gymnocephalus schraetzer*) and the Danube ruffe (*Gymnocephalus baloni*) do feature in schedules of the Berne Convention.

Under the European Union (EU) Habitats Directive (Council Directive 92/43/EEC on the Conservation of Natural Habitats and of Wild Fauna and Flora), the European ruffe is also not mentioned. However, the schraezter is listed in two Annexes as an indicator of habitats of concern.

European ruffe are, however, variously listed as potential pests given their propensity to threaten populations of native fish in places where they have been introduced beyond their native range.

Balancing these negative impacts, ruffe also have potential indirect nature conservation benefits. There is a suggestion that ruffe may predate upon juvenile American signal crayfish (*Pacifastacus leniusculus*). These crustaceans have been widely introduced around the world, including across much of Britain and continental Europe, causing substantial ecological and potential socio-economic damage. If ruffe can help suppress signal crayfish numbers by grazing on juveniles, they may well serve a useful nature conservation outcome.

Whatever their specific inclusion under nature conservation designations, or rather their lack of inclusion, ruffe nonetheless act as important elements of the ecosystems in which they occur, feeding on small animal matter and in turn serving as food for chains of predatory animals. Playing important roles as links in food webs, ruffe in their natural habitats make significant contributions to the integrity and health of freshwater ecosystems. Their conservation consequently matters for more than altruistic reasons, including for the diverse benefits that these ecosystems confer upon humanity.

PET RUFFE

Günther Sterba's 1962 book *Freshwater Fishes of the World* was something of a definitive text of its time on aquarium fishes. In consideration of food for small perch kept in unheated aquaria, Sterba recommended:

> . . . *midge and other insect-larvae, worms and slugs; larger individuals will also take young fishes.*

Sterba then continued to note that 'Ruff' [sic] and 'Schraetzer':

> . . . *can also be kept successfully for a long time under the conditions just described.*

RUFFE SOCIETIES

Unlike pike, roach, carp, tench, perch, barbel and even gudgeon, ruffe do not have a dedicated society. Or, at least, not yet, though there are discussions taking place about forming one as this book is being written.

There are also virtual digital communities in the shape of members of at least two Facebook social media pages.

A RUFFE STORY

I have struggled somewhat to find a published story specifically about ruffe, as distinct from stories such as those of Chekhov and other writers who refer to ruffe tangentially.

The dedicated ruffe story that I did find was titled *The Great Ruffe Hunt*. This tale is to be found in my own 2008 book *The Little Book of Little Fishes*. The story relates, literally, to a 'great ruffe hunt' in a Westcountry canal involving me, my good angling buddy Sid and my then eight-year-old daughter. Unfortunately, this story is a bit long to reproduce here in its entirety, so I'll have to leave you to track down a copy of this book that is badly in need of being reprinted!

RUFFE: A FISH OF MYTH AND LEGEND!

OK, so maybe I am overstating the 'fish of myth and legend' case somewhat!

The important message here, though, is that there is rather more depth and breadth to the humble ruffe than might be first suspected in terms of its contributions to humanity.

But, as lovers of ruffe, we probably suspected that all along!

NOTES

1 Various other purportedly authoritative sources give a Latin name of *G. cernuus*, wrongly reproduced in various other publications, but *G. cernua* is the correct name given in the definitive *Eschmeyer's Catalog of Fishes*.

2 For those that want to know, broth is a savoury liquid resulting from simmering bones, meat or vegetables, whereas soup is a hot or cold liquid food made of combined ingredients.

3 The full citation for Svanberg and Locker (2020) is in the *Ruffe bibliography* at the end of this book.

The following works are referenced in this book, with my thanks to the authors concerned where quoted.

Adams, C.E. and Maitland, P.S. (1998). The Ruffe Population of Loch Lomond, Scotland: Its Introduction, Population Expansion, and Interaction with Native Species. *Journal of Great Lakes Research*, 24(2), pp. 249–262. DOI: https://doi.org/10.1016/S0380-1330(98)70817-2.

Anon. (1846). *The Art of Angling Greatly Improved: Containing the Most Esteemed Methods of Angling for Pond and River Fish; The Baits for Each, and How to Obtain and Preserve Them the Choosing of Rods and Tackle; Also, Instructions in Every Branch of Fly-Fishing, and for the Making of Flies.* Thomas Richardson And Son, Derby.

Bell, Arthur P. (1926). *Fresh-Water Fishing for the Beginner.* Warne's Recreation Books, Frederick Warne & Co. Ltd., London.

Berners, Dame Juliana. (1496). *Treatyse of Fysshynge with an Angle.* Wynkyn de Worde of Westminster, London.

California Academy of Sciences. (2021). *Eschmeyer's Catalog of Fishes* (online version, updated 1st March 2021). Institute for Biodiversity Science and Sustainability, California Academy of Sciences, San Francisco.

Canal and River Trust. (Undated). It's a Ruffe Old World Out There. Canal and River Trust. (https://canalrivertrust.org.uk/enjoy-the-waterways/fishing/related-articles/the-fisheries-and-angling-team/its-a-ruffe-old-world-out-there.)

Cholmondeley-Pennell, H. (1863). *The Angler-Naturalist.* George Routledge and Sons, London.

Cone, D., Eurell, T. and Beasley, V. (1994). A Report of *Dactylogyrus Amphibothrium* (Monogenea) on the Gills of European Ruffe in Western Lake Superior. *The Journal of Parasitology*, 80(3), pp. 476–478. DOI: https://doi.org/10.2307/3283421.

Coxon, Henry. (1896). *A Modern Treatise on Practical Coarse Fish Angling: How to Catch Fish.* Charles H. Richards, Nottingham. (Republished in 2004 by The Medlar Press, Ellesmere).

Everard, Mark. (2008). *The Little Book of Little Fishes.* The Medlar Press, Ellesmere.

Everard, Mark. (2020). *The Complex Lives of British Freshwater Fishes.* CRC/Taylor and Francis, Boca Raton and London.

Francis, Francis. (1867). *A Book on Angling.* Longmans, Green & Co., London.

Franck, Richard. (1694). *Northern Memoirs.* Henry Mortclock, London.

Garnett, Dominic and Mytton, Andy. (2019). *Hooked on Lure Fishing.* Merlin Unwin Books Ltd, Ludlow.

Giles, Nick. (1994). *Freshwater Fish of the British Isles: A Guide for Anglers and Naturalists.* Swan Hill Press, Shrewsbury.

Gutsch, M. and Hoffman, J. (2016). A Review of Ruffe (*Gymnocephalus Cernua*) Life History in Its Native versus Non-Native Range. *Reviews in Fish Biology and Fisheries,* 26, pp. 213–233. DOI: https://doi.org/10.1007/s11160-016-9422-5.

Houghton, W., MA, FLS, The Reverend. (1879). *British Fresh-Water Fishes.* William Mackenzie, London. (Note: This book has been reprinted over the decades by numerous publishers, for example by The Peerage Press, London, in 1981.)

Mansfield, Kenneth. (1958). *Small Fry and Bait Fishes: How to Catch Them.* Herbert Jenkins, London.

Marshall-Hardy, Eric. (1943). *Coarse Fish.* Herbert Jenkins Limited, London.

Muus, Bent J. and Dahlstrom, Preben. (1967). *Collins Guide to the Freshwater Fishes of Britain and Europe.* Collins, London.

Pinder, A.C. (2001). *Keys to Larval and Juvenile Stages of Coarse Fishes from Fresh Waters in the British Isles.* Freshwater Biological Association Scientific Publications Volume 60. Freshwater Biological Association, Windermere.

Ransome, Arthur. (1929). *Rod and Line.* Jonathan Cape, London.

Sheringham, Hugh Tempest and Studdy, G.E. (1925). *Fishing: Its Cause, Treatment and Cure.* Philip Allan and Co., London.

Stepien, A., Dillon, A.K. and Chandler, M.D. (1998). Genetic Identity, Phylogeography, and Systematics of Ruffe *Gymnocephalus* in the North American Great Lakes and Eurasia. *Journal of Great Lakes Research,* 24(2), pp. 361–378. DOI: https://doi.org/10.1016/S0380-1330(98)70827-5.

Sterba, Günther. (1962). *Freshwater Fishes of the World.* Vista Books, London.

Svanberg, I. and Locker, A. (2020). Caviar, Soup and Other Dishes Made of Eurasian Ruffe, *Gymnocephalus Cernua* (Linnaeus, 1758): Forgotten Foodstuff in Central, North and West Europe and Its Possible Revival. *Journal of Ethnic Foods,* 7(3). DOI: https://doi.org/10.1186/s42779-019-0042-2.

Walton, Izaak and Cotton, Charles. (1653). *The Compleat Angler*. Maurice Clark, London. (Available these days in many editions and from various publishers).

Wells, A. Lawrence. (1941). *The Observer's Book of Freshwater Fishes of the British Isles*. Frederick Warne and Co. Ltd, London.

Wheeler, Alwyne. (1969). *The Fishes of the British Isles and North-West Europe*. Macmillan, London.

Professor Mark Everard is a scientist, author and broadcaster, working on water and ecosystems around the world. He also has an irrationally large interest in fish!

Mark is also a passionate angler, getting out whenever he can after coarse, game and sea fish. He has an enviable track record of specimen fish with a particular passion for roach, dace and mahseer, and has long been a champion of 'little fishes' such as the ruffe.

Mark Everard is often referred to as Dr Redfin in the angling press for his special passion for roach.

For more about Mark and his work, see www.markeverard. co.uk.